VITAL INFO

"Nuts & Bolts" of Life-Science

Author: Dr. J. Jarrod Dunning

1st Edition

Copyright © 2018 Dr. J. Jarrod Dunning

All Rights Reserved

Dedication

Principal Davida Hill-Johnson

Malachi Wilkerson Middle School

Thank you for your help and support. Your leadership is second to none. Under your leadership we were able to develop a science curriculum that is revolutionizing the way science teachers approach teaching and learning science.

Again thank you Principal Davida Hill-Johnson.

Romans 16:27
May Glory Be To God, Who Alone Is Wise.

Table of Contents

How to Use Vital Info 1

Exploration 1: Cell Theory 5

Exploration 2: Cells Eukaryote & Prokaryote 7

Exploration 3: Heterotroph & Autotroph Organelles 9

Exploration 4: Biological Hierarchy 13

Exploration 5: Ecosystem 15

Exploration 6: Carrying Capacity 19

Exploration 7: Eco-Succession... 21

Exploration 8: Symbiotic Relationships 23

Exploration 9: Water Purification 25

Exploration 10: Autotrophs 27

Exploration 11: Epigenetics 31

Exploration 12: Inheritance & Variations 33

Exploration 13: Gene Mutation.. 37

Exploration 14: Genetic Engineering 39

Exploration 15: Fossil Record 41

Exploration 16: Evolutionary Decent.................................. 43

Exploration 17: Vestigial Structures.................................... 45

Exploration 18: Natural Selection 47

Exploration 19: Nature of Science 49

How to Use Vital Info

Life Science

Firstly, thank you for considering Vital Info as a supplemental educational tool to learning Life Science. Please enjoy the explorations and engaging methods of learning Life Science. Utilize Vital Info throughout the school year to activate background knowledge when teaching Life Science. Vital Info can be helpful when writing lesson plans. Lesson plans created with the aid of Vital Info bridge the learning gap so that all students can find success in learning Life Science.

The purpose of Vital Info is to increase students' level of scientific rigor & life science comprehension through metacognition. Metacognition is the process of thinking about thinking. Vital Info is designed to share my knowledge, skills and abilities in how I am teaching and learning life science. Also, I would like students, parents and educators to use Vital Info to gain a better understanding of life science.

This book consists of four highlights: Life Science Explorations, Learning Targets, Mnemonic Devices and Opportunities to Respond.

Also, I use strategies such as using Whole Brain Learning and Graphic Organizers to break down each lesson.

- The **science explorations** are unpacked so students and educators can analyze essential questions, teach and learn.

- The **learning targets** are what students will be able to analyze once they have mastered the exploration.

- **Mnemonic devices** is a learning tool used to increase students' retention of the lesson taught; a unique breakdown of scientific terms, processes and concepts.

- **Opportunities to respond (OTR)** allow students to explore science using critical thinking skills while increasing analyzation skills to answer the essential questions.

- **Whole brain learning** is a learning tool used to increase student engagement. Students incorporate their bodies to learn key scientific concepts.

- **Graphic organizers** are a great way for students to make learning connections and comprehend the lesson in an organized creative way.

I recommend that readers review all explorations and insert their own creative methods of teaching and learning according to the cultural capital of students. Cultural capital is the knowledge and experience a student carries to school while learning. I encourage students and educators to take Vital Info into class and use it as a necessary go-to referential guide to Life Science.

Vital Info is undoubtedly the "Nuts & Bolts" of Life Science. Some science topics may not be covered in this book. I recommend incorporating YouTube to access additional Life Science resources.

Author,

Dr. J. Jarrod Dunning

Dr. J. Jarrod Dunning

Exploration 1

Exploration 1: Engage in arguments from evidence to support claims of the cell theory.

Students will:

- Identify the presence or absence of cells in living and nonliving things.
- Identify the presence of cells in a variety of organisms, including unicellular and multicellular organisms.
- Identify different types of cells within one multicellular organisms.

Move into the Know

Essential Question: How can you analyze the cell theory?

What are the three concepts of the Cell Theory?
Answer:
1. Cells are the basic unit and structure for life.
2. Cells come from pre-existing cells.
3. All organisms are composed of one or more cells.

What two scientists argued about cells deriving from pre-existing cells?
Answer:
1. Remake and Rudolph Virchow.
2. Virchow took Remake's idea regarding cells coming from pre-existing cells.

What scientist created the Microscope, and who did he argue with?
Answer:
1. Robert Hooke discovered the microscope.
2. Robert Hooke told Isaac Newton that he was inspired by his work.
3. Isaac Newton told Robert Hooke "I don't know you."

Which other scientists influenced the cell theory?
Answer:
1. Matthias Schleiden and Theodor Schwann

Exploration 2

Exploration 2: Gather and synthesize information to explain how prokaryotic and eukaryotic cells differ in structure and function, including the methods of asexual reproduction.

Students will:

- Analyze how prokaryotic cells are microscopic.
- Identify prokaryotes, including the bacteria and cyanobacteria.
- Describe the function of prokaryote and eukaryote cells.

Move into the Know

Essential Question: How can you analyze cells?

What is prokaryote & eukaryote cells?
Answer:
1. Prokaryote cells are *single cellular*. Another word for single cellular is *unicellular*.
2. Prokaryote cells do not have a nucleus. Prokaryote cells do not have a cell membrane.
3. **Example** of a Prokaryote cell is bacteria.
4. Eukaryote cells are *multicellular*. Multicellular means *many cells*.
5. Eukaryote cells do have a nucleus. Eukaryote cells do have a cell membrane.

Vital Info

6. **Example** of a Eukaryote cell are *autotrophs* (plants) & *heterotrophs* (animals).

Define **Meiosis**.
Answer: Meiosis is the fertilization of gametes.

What are **gametes**?
Answer: Gamete is sperm & egg. Gametes are *haploid* cells which mean the sperm contains 23 chromosomes and the egg contains 23 chromosomes.

Explain a Diploid Cell.
Answer: Once meiosis is complete the cells are called *diploid*, which means the cell contains 46 chromosomes.

The 23rd chromosome is a sex-linked chromosome. True or False
Answer: True
Explanation: The 23rd chromosome determines the gender of the cell. **XX=Female** and **XY=Male**.

What is **Mitosis**?
Answer: Mitosis is Cell Division, which undergoes the following processes: **Interphase**, **Prophase**, **Metaphase**, **Anaphase** and **Telophase (I.P.M.A.T.).**

What is the mnemonic device we use for Cell Division?
Answer: I.P.M.A.T.

Explain Cytokinesis.
Answer: The final step of mitosis is called *cytokinesis*, which means the separation of one cell into two cells.

Exploration 3

Exploration 3: Construct an explanation of the function of specific cell structures for maintaining a stable environment.

Students will:

- Describe the function of organelles
- Identify roles of organelles in maintaining a stable environment.
- Analyze key differences between animal and plant cells.

Move into the Know

Essential Question: How can you analyze animal and plant organelles?

How many animal organelles are there?
Answer: There are **13** animal organelles.

Mnemonic device used for animal organelles:
Nate Ran Round Pretty Slow Getting Lemonade Minute Maid Coffee Pineapples Mango & Fritos

Nate (Nucleus)	Ran (Ribosome)
Round (Rough Er)	Pretty (Plasma Membrane)
Slow (Smooth Er)	Getting (Golgi Apparatus)
Lemonade (Lysosome)	Minute (Microtubules)
Maid (Microfilament)	Coffee (Centrioles)
Pineapples (Peroxisome)	Mango (Mitochondria)
Fritos (Flagella)	

Vital Info

What are the plant organelle **Three C's**?
Answer:
Cell Wall, **C**hloroplast & **C**entral Vacuoles.

How many plant organelles are there? (Include the animal organelles)
Answer: There are **16** plant organelles.

What are the functions of the Animal and Plant organelles?
Animal organelles:
Nucleus - Control Center
Ribosome - Makes Proteins
Rough Er - Transport Proteins
Plasma Membrane - Allows materials to move in and out **Smooth Er** - Stores Lipids
Golgi Apparatus - Package Center
Lysosome - Eliminates Waste
Microtubules - Provides Structure
Microfilament - Provides Structure
Centrioles - Cell Division
Peroxisome - Detox
Mitochondria - Power House
Flagella - Cell Movement

<u>Add</u> the below **Three C's** to create a **plant organelle**.
Cell Wall- Rigid Structure **Chloroplast**-Light Dependent Reactions (LDR) **Central Vacuole**-Eliminates waste

What does the chloroplast contain?
Answer: Chloroplast contains disc-like structures called *thylakoids*.

What are the thylakoids surrounded by?
Answer: The thylakoids are surrounded by a fluid-filled space called the *stroma*.

Where is the chlorophyll molecule located?
Answer: The chlorophyll molecule is located inside the thylakoids.

What reactions take place inside the chlorophyll molecule?
Answer: The light dependent reactions take place inside the chlorophyll molecule.

What are light dependent reactions?
Answer: Light dependent reactions is the conversion of red and blue light into energy called ATP.

What is Photosynthesis?
Answer: Photosynthesis is the manufacture of glucose.

What does cellular respiration begin with?
Answer: Cellular respiration begins with glucose.

Where does cellular respiration take place in the cell?
Answer: Cellular respiration takes place in the mitochondria.

How can you apply cellular respiration to animals?
Answer: Cellular respiration in animals requires eating in order to gain energy.

What cycle takes place in the stroma?
Answer: The Calvin cycle takes place in the Stroma.

What is the stroma?
Answer: The Stroma is a fluid filled space.

Explain the Calvin cycle.
Answer: The Calvin cycle has three stages. The first stage is **carbon fixation**; second stage is **reduction**; third stage is **regeneration**.

How many times will the Calvin cycle rotate to produce one molecule of glucose?
Answer: The Calvin cycle will rotate six times to produce one molecule of glucose.

Exploration 4

Exploration 4: Construct models and representations of organ systems to demonstrate how multiple-interacting organs and systems work together to accomplish specific functions.

Students will:

- Analyze Biological Hierarchy.

- Identify specialized cells that make up specialized tissues.

- Identify specialized tissues that make up organs.

- Identify major organs of the body systems.

Move into the Know

Essential Question: How can you analyze Biological Hierarchy?

Mnemonic Device: *AM+CTOOO+PC+ EBB*

What is the Biological Hierarchy? (*Read from left to right*)

Answer:

Atoms	**M**olecules	**C**ells
Tissue	**O**rgans	**O**rganisms
Organ System	**P**opulation	**C**ommunity
Ecosystem	**B**iomes	**B**iosphere

What is a population?
Answer: A population is one species.
Example: 15 Lions.

What is a community?
Answer: A community is more than one species.
Example: Five lions, four tigers and ten bears.

What is an Ecosystem?
Answer: An ecosystem is the symbiotic relationship of how organisms support living things within an environment.

What is Ecology?
Answer: Ecology is the study of how organisms interact within an environment.

What are Biomes?
Answer: A biome is a community of plants and animals that has common characteristics for the environment it exists in.

Mnemonic Device: Biomes 4T's DoG

4T's: **T**undra: **T**aiga: **T**ropical Rainforest: **T**emperate Forest **(North America)**
Dessert: **G**rassland

Exploration 5

Exploration 5: Examine the cycling of matter between abiotic and biotic parts of the ecosystem to explain the flow of energy and the conservation of matter.

Students will:

- Describe how organisms can be classified as producers, consumers, and/or decomposers.
- Describe how abiotic parts of an ecosystem can provide matter to biotic organisms.
- Identify how biotic organisms of an ecosystem can provide matter to abiotic parts.

Move into the Know

Essential Question: How can you analyze the ecosystem?

What does **abiotic** factors mean?
Answer: Nonliving things.

What does **biotic** factors mean?
Answer: Living things.

What is the Energy Pyramid?
Answer: The Energy Pyramid shows the **flow of energy** as it passes from one **Trophic Level** to the next.

Primary Producer	Primary Consumer
Plants	*Rabbits / Insects*
Gymno/Angiosperms	**Herbivores**

Secondary Consumer	Tertiary Consumer	Apex
Snakes / Frogs	*Hawks*	
Carnivores	**Top Carnivores**	**(predators)**

What models are used to explain the flow of energy within an ecosystem?

Answer: The models used to explain the flow of energy within an ecosystem are the **Energy Pyramid**, **Food Web** and **Food Chain**.

What is a Food Web?

Answer: A food web is a model that shows an **overlapping food chain**.

What is a Food Chain?

Answer: A food chain shows the **flow of energy** as **one species eats another**.

What are the **bio-geochemical** cycles?

Answer: Hydrogen Cycle, Nitrogen Cycle, Carbon Cycle & Phosphorus Cycle are biogeochemical cycles.

- Hydrogen Cycle

Mnemonic Device: (P.C. R.I.I.P. D.T.E.C.)
1. **P**recipitation / **C**ollection
2. **R**unoff/**I**nterception /**I**nfiltration /**P**ercolation
3. **D**ischarge/**T**ranspiration/**E**vaporation/ **C**ondensation

What is transpiration?

Answer: Transpiration is when water and nutrients are absorbed by autotrophs from the roots, to the stems, to the leaves, and finally oxygen is released into the atmosphere.

- Nitrogen Cycle

Mnemonic Device: (D.A.N.N.N.)
1. **D**enitrification
2. **A**mmonification
3. **N**itrification
4. **N**itrogen Assimilation
5. **N**itrogen Fixation

What is the percentage of nitrogen in the air?
Answer: 78% of nitrogen is in the air.

What is the percentage of oxygen floating in the air?
Answer: 20% of oxygen is floating in the air.

What is nitrification?
Answer: Nitrification is when **ammonia** (NH3) is converted into **nitrites** (NO2) and then into **nitrates** (NO3).

What process does aquaponics use?
Answer: Nitrification.

What is aquaponics?
Answer: Aquaponics is the combination of hydroponics and aquaculture.

What is hydroponics?
Answer: Hydroponics is the cultivation of plants without soil.

What is aquaculture?
Answer: Aquaculture is the cultivation of fish.

- Carbon Cycle

In what ways can carbon enter and leave the atmosphere?
Answer: Autotrophs take carbon out and heterotrophs put carbon in the atmosphere.

- Phosphorus Cycle

In what ways can phosphorus enter the atmosphere?
Answer: Phosphorus is secreted from the **lithosphere** which is located in the **earth's crust**. Autotrophs love phosphorus.

Exploration 6

Exploration 6: Analyze and interpret data to provide evidence regarding how resource availability impacts individual organisms and populations of organisms within an ecosystem.

Students will:

- Describe how, in all ecosystems, organisms and populations with similar requirements for food, water, oxygen, or other resources, may compete with each other for limited resources.

- Analyze the carrying capacity of an ecosystem.

Exploration 7

Exploration 7: Use empirical evidence from patterns and data to demonstrate how changes to physical or biological components of an ecosystem can lead shifts in populations.

Students will:

- Describe how ecosystems are dynamic in nature and can change over time.
- Analyze how disruptions to any physical or biological component of an ecosystem can lead to shifts in all its' populations.
- Identify how changes in the physical or biological components of an ecosystem, and determine how these changes can lead to shifts in populations of species.

Move into the Know

Essential Question: How can you analyze Eco-Succession?

What is Primary Succession & Secondary Succession?

Answer: Succession is how things grow back over time due to minor and major disturbances.

Primary Succession ***Bare Rock*** Ex. **Meteorite**_(Major Disturbance)

Secondary Succession ***Soil Remain*** Ex. **Wild Fire** (Minor Disturbance)

Exploration 8

Exploration 8: Construct an explanation to predict patterns of interactions in different ecosystems concerning relationships among organisms.

Students will:

- Analyze how competitive relationships occur when organisms within an ecosystem compete for shared resources.

- Identify how predatory interactions occur between organisms within an ecosystem.

- Identify how mutually-beneficial interactions occur between organisms within an ecosystem.

- Analyze how some organisms are so dependent upon one another that they cannot survive alone.

Move into the Know

Essential Question: How can you analyze symbiotic relationships?

What is symbiosis?
Answer: Interaction between two different organisms living in close physical association.

What are the **three symbiotic relationships** and functions of each?

Answer:
1. **Commensalism** *One species benefits while the other is not affected* **Ex.** Shark with smaller fish alongside
2. **Mutualism** *Both species benefit* **Ex.** Aquaponics
3. **Parasitism** *One species benefits while the other is harmed* **Ex.** Mosquito Bite

Exploration 9

Exploration 9: Engage in an argument to defend the effectiveness of a design solution that maintains biodiversity and ecosystem services.

Students will:

- Identify evidence about a performance of a given design solution.

- Analyze how biodiversity is a variety of species found in the earth's ecosystems.

- Identify how the completeness of the biodiversity of an ecosystem is often used as a measure of health.

- Analyze how changes in biodiversity can influence human resources and ecosystem services.

Move into the Know

Essential Question: How can you analyze water purification?

What is the pH range?
Answer: 0 – 14

What is the pH range of 0-7?
Answer: Acid

What is the pH of 7?
Answer: Neutral

What is the pH range of 7-14?
Answer: Base

What is the pH of tap water?
Answer: 7.0

What is the pH range for healthy water?
Answer: 6.5 - 8.5

What is the pH range for plants?
Answer: 5.5 - 6.5

What is Reverse Osmosis?
Answer: RO is a water purification system that uses a semipermeable membrane to remove large particles.

What are some designed solutions that maintains biodiversity?
Answer: Reverse Osmosis, Hydroponics and **Aquaponics** are a few designed solutions that maintain biodiversity through sustainability.

Exploration 10

Exploration 10: Use evidence and scientific reasoning to explain how characteristic animal behaviors and specialized plant structures affect the probability of successful reproduction of both animals and plants.

Students will:

- Identify the Sepal and Corolla.
- Analyze how angiosperms are Flowers and Fruits.
- Analyze how gymnosperms are Pine trees, which produce Pine cones called "a naked seed".

Move into the Know

Essential Question: How can you analyze Plants?

What is another word for plant?
Answer: Autotroph

What are the three types of plants?
Answer: Algae, Nonvascular plants and Vascular plants.
What is the difference between **Nonvascular** plants and **Vascular** plants?
Answer:
Vascular Plants
Have Roots, Stem & Leaves *Ex. Banana Peppers*
NonVascular Plants
No Roots, No Stem & No Leaves *Ex. Mosses*

Mnemonic Device: Plant Parts (*Music Stays Alive Forever & Funny People Sing Songs*)

What are Angiosperms?
Answer: Angiosperms are **flowers** and **fruits.**

What are Gymnosperms?
Answer: Gymnosperms are Pine trees which produce Pine cones called naked seeds.

What are the parts of a plant?
Answer: Male part of an autotroph: Stamen, Anther & Filament.
Female part of an autotroph: Pistil, Stigma & Style.

What does the Anther produce?
Answer: Pollen

What is the function of the Stigma?
Answer: The stigma is **sticky**. The stigma is located at the top of the Pistil. The pollen produced by the anthers stick to the stigma to pollinate the flower.

Where do plants absorb water and nutrients?
Answer: Plants absorb water and nutrients in their roots through **osmosis** and **diffusion**.

What is the difference between osmosis and diffusion?
Answer: Osmosis is the process plants use to absorb **water**. Diffusion is the process plants use to absorb **nutrients**.

What is another word for Corolla?
Answer: Another word for Corolla is *petal*.

What two colors do plants absorb?
Answer: Red & Blue

What does the xylem and the phloem transport?
Answer: The **xylem** transports **water** and the **phloem** transports **glucose**.

What do plants produce?
Answer: Plants produce **oxygen** through a process called Transpiration.

What process is used for plants to grow their own food?
Answer: Photosynthesis is the process plants use to make glucose which is sugar.

What is **cellular respiration**?
Answer: Plants use food (**glucose**) to make energy in the form of ATP.

What is **germination**?
Answer: Germination is the process in which a seed begins to sprout and tropisms can be observed.

What is **Tropism**?
Answer: Tropism is the movement of a plant toward a stimulus.

What is **Phototropism**?
Answer: The movement of a plant towards light.

What is **Thigmotropism**?
Answer: The curling movement of a plant on a solid object.

What is **Chemotropism**?
Answer: The movement of a plant towards chemicals (**Nutrients**).

What is **Gravitropism**?
Answer: The movement of a plant towards gravity. Negative gravitropism is the stem growing up. Positive gravitropism is the roots growing down.

What is the difference between **asexual** and **sexual reproduction**?
Answer:
1. *Asexual* reproduction: the **offspring are identical**. Ex. Plants
2. *Sexual* reproduction: the **offspring are not identical**. Ex. Animals

Exploration 11

Exploration 11: Analyze and interpret data to predict how environmental conditions influence the growth of organisms.

Students will:

- Identify how environmental factors can influence growth.

- Identify how genetic factors can influence growth.

- Analyze how changes in the growth of organisms can occur as specific environmental and genetic factors change.

- Analyze how epigenetics are conditions that influence the growth of organisms.

Exploration 12

Exploration 12: Construct and use models to explain that genetic variations between parent and offspring occur asa result of genetic differences in randomly inherited genes located on chromosomes and that additional variations may arise from alteration of genetic information.

Students will:

- Analyze how chromosomes are the source of genetic information.

- Analyze how organisms reproduce, either sexually or asexually and transfer their genetic information to offspring.

- Identify variations of inherited traits from parent to offspring, and analyze how these traits derive from the genetic differences of chromosomes inherited.

Move into the Know

Essential Question: How can you analyze Inheritance and Variation?

What model can you use to explain genetic differences?
Answer: Punnett Square

What is inheritance?
Answer: Passed on traits from parent to offspring.

What is a change or alteration in DNA (**De-oxy-ribo-nucleic-acid**)?
Answer: Mutation

What are the four DNA nucleotides?
Answer:
1. Adenine
2. **Thymine**
3. Cytosine
4. Guanine

What are the four RNA nucleotides?
Answer:
1. Adenine
2. **Uracil**
3. Cytosine
4. Guanine

What does a DNA molecule look like?
Answer: The DNA molecule is a double-helix which looks like a twisted ladder.

What is the genetic process to spread cells?
Answer:
1. Replication (**DNA is Copied**)
2. Transcription (Transferring information from **DNA** to **RNA**)
3. Translation (**Manufacture of Protein**)

What is MRNA and TRNA?
Answer:
1. Messenger RNA (**Ribo-nucleic-acid**)
2. Transfer RNA (**Ribo-nucleic-acid**)

What is the difference between Heterozygous and Homozygous traits?
Answer: Heterozygous have different traits and homozygous have the same traits.

Exploration 13

Exploration 13: Construct an explanation from evidence to describe how genetic mutations result in harmful, beneficial or neutral effects to the structure and function of an organism.

Students will:

- Identify genes that are located in the chromosomes of cells.

- Analyze how each chromosome pair contains two variants of genes.

- Describe how genes control the production of proteins.

- Describe how proteins affect the structures and functions of the organism, thus changing traits.

Move into the Know

Essential Question: How can you analyze the results of Gene Mutations?

Mnemonic Device: Bright House Network

What are the three results of a gene mutation?
Answer:
1. *Beneficial* (Gene mutations are Rare)
2. *Harmful* (Most Gene mutations are harmful)
3. *Neutral* (Many Gene mutations are neutral)

Vital Info

What is a **change/alteration** of **De-oxy-ribo-nucleic Acid (DNA)?**
Answer: An **alteration** in **DNA** is called a **mutation**.

What process do genes use to manufacture protein?
Answer: Translation

What is Gene Mutation?
Mnemonic Device: S.I.D.
Answer:
1. Substitution (*Wrong Base Pairs*) **Ex. Adenine - Cytosine**
2. Insertion (*Three Codons insertion of a 4th*) **AUG to AUCG**
3. Deletion (*Three Codons deletion of one*) **Ex. AUG to A_G**

What is Chromosomal Mutation?
Mnemonic Device: D.D.I.T. (Dogs Dig In Trash)
Answer:
1. Deletion
2. Duplication
3. Inversion
4. Translocation

Exploration 14

Exploration 14: Gather and synthesize information regarding the impact of technologies on the inheritance and/or appearance of desired traits in organisms.

Students will:

- Analyze and interpret data through technologies, humans have the capacity to influence certain characteristics of organisms.

- Analyze how a human can choose desired parental traits determined by genes, which are then passed to offspring.

Move into the Know

Essential Question: How can you analyze Genetic Engineering?

What is another word for genetic engineering?
Answer: Genetic modification.

What is genetic modification?
Answer: The direct manipulation of an organism's genes using biotechnology.

What is gene therapy?
Answer: Gene therapy is the therapeutic delivery of nucleic acid into a patient's cells as a drug to treat disease instead of using drugs or surgery.

Exploration 15

Exploration 15: Analyze and interpret data for patterns of change in anatomical structures of organisms using the fossil record and the chronological order of fossil appearance in rock layers.

Students will:

- Analyze how the **law of super position** is when **older rock layers** are located on the bottom and the **younger rock layers** are located near the top.

- Organize the given data in a way that allows for the identification and analysis of similarities and differences. Life evolved from simple to more complex forms of life.

- Analyze how the **evidence of evolution** is found in the **fossil record**.

Exploration 16

Exploration 16: Construct an explanation based on evidence for the anatomical similarities and differences between modern and fossil organisms, including living fossils.

Students will:

- Describe how anatomical similarities and differences among organisms can be used to infer evolutionary relationships among modern organisms and fossil organisms.

Exploration 17

Exploration 17: Obtain and evaluate pictorial data to compare patterns in the embryological development across multiple species to identify relationships not evident in the adult anatomy.

Students will:

- Analyze pictorial data and compare patterns in the embryological development across multiple species.

- Analyze how **vestigial structures** are structures your ancestors used that are no longer needed in modern organisms.

Exploration 18

Exploration 18: Construct an argument from evidence that shows that natural selection acting over generations may lead to the predominance of certain traits that support successful survival and reproduction of a populations, which in turn suppresses other traits.

Students will:

- Articulate a statement that relates a given phenomenon to a scientific idea, including natural selection and traits.

- Identify and use multiple valid and reliable sources of evidence to construct an explanation for natural selection and its effect on traits in a population.

- Use reasoning to connect the evidence and support an explanation for natural selection and its effect on trait in a population.

Move into the Know

Essential Question: How can you analyze natural selection?

Mnemonic Device: O.V.A.S.

1. **Overproducing**
   Species must overproduce to overcome diseases and predators.

2. **Variation**
   Species have a variety of phenotypes.
3. **Adaptation**
   Any inherited trait that gives an offspring the advantage within an environment.
4. **Selection**
   Understand that you cannot get selection without adaptation.

Exploration 19

Exploration 19: Nature of Science. Scientist use the scientific method in order to conduct scientific investigations.

Move into the Know

Essential Question: How can you analyze the Nature of Science?

Who uses the scientific method to conduct scientific investigations?

Answer: Scientist use the scientific method in order to conduct scientific investigations.

What do you need to create a graph?

Answer: In order to create a graph, you need *three titles: identify the independent and dependent variables, draw conclusions and make predictions based on the patterns of the data.

*Independent variable (**x-axis**) title, dependent variable (**y-axis**) title and the title of the **entire graph.

****Combine** the **x-axis title** and the **y-axis title** in order to create the **title** for the **entire graph**.

Which axis is the independent variable located?
Answer: The independent variable is located on the **x-axis**.

Which axis is the dependent variable located?
Answer: The dependent variable is located on the **y-axis**.

What is the Scientific Method?
Answer: Choose a Problem →Research that Problem→ Develop a Hypothesis Design an Experiment → Test your Hypothesis → Organize your Data → Draw a Conclusion

Contact

Dr. J. Jarrod Dunning

Facebook: Nutz Boltz

Youtube: Dr. J. Jarrod Dunning

Email: jarroddunning@gmail.com

www.ingramcontent.com/pod-product-compliance
Lightning Source LLC
Chambersburg PA
CBHW030052230526
45471CB00003B/1059